Sea Dragons and Rainbow Runners

Exploring Fish with Children

BY SUZANNE SAMSON

ILLUSTRATED BY PRESTON NEEL

ROBERTS RINEHART PUBLISHERS

While hunting for treasure at sea, I discovered a Rainbow Runner searching for gold. I decided to follow her.

On my journey I saw…

A Sawfish creating a path, and…

A Soapfish submerged in a bath, and…

A Trunkfish commencing a trip, and…

A Sea Dragon confronting a ship, and…

A Hammerhead employed as a tool, and…

Two Swordfish engaged in a duel, and…

A Sunfish acquiring a tan, and...

A Spanish Dancer twisting her fan, and...

Christmas Trees with lights that glow, and...

Seahorses in a rodeo, and…

A Boxer Crab throwing a punch, and…

A Fat Sleeper dreaming of lunch, and…

A Cowfish producing some milk, and…

Jellyfish floating like silk, and…

A Clown fish leaping through hoops, and…

A Sergeant Major drilling troops, and…

A Trumpetfish joining a band, and…

A Flyingfish gliding towards land, and…

A Dogfish utilizing his nose, and…

A Frogfish kissing a princess, and…

A Sailfish with one sailor less, and…

A Schoolmaster teaching a class, and… …

A Footballer Trout completing a pass.

The Rainbow Runner eventually led me to a treasure chest full of gold. Unfortunately, a Pirate Perch had beaten us to it. So, I stored my memories of the spectacular life at sea, and that was treasure enough for me.

Rainbow Runner
Elagatis bipinnulata
Can be found over reefs and near the surface of deep water.
Grows to about 4' (1.2m).

Smalltooth Sawfish
Pristis pectinata
Shallow areas of coastal waters, estuaries and large rivers.
Grows to about 18' (5.5m).

Greater Soapfish
Rypticus saponaceus
In shallow water over reefs and rocks.
Grows to about 13" (33 cm).

Scalloped Hammerhead
Sphyrna lewini
Open seas near the surface, have been found in estuaries.
May grow to over 13 1/2' (4.1 m).

Spotted Trunkfish
Lactophrys bicaudalis
Coral reefs; grassy areas in shallow waters.
Grows to about 21" (53 cm).

Leafy Sea Dragon
Phycodurus eques
Kelp areas along the southern coast of Australia.
Grows to about 10" (25 cm).

Swordfish
Xiphias gladius
Open seas.
Grows to about 15' (4.6 m).

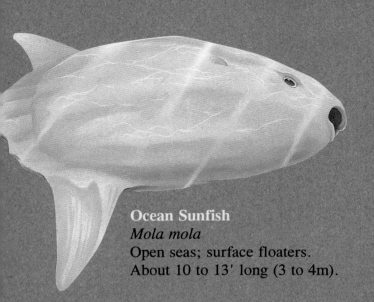

Ocean Sunfish
Mola mola
Open seas; surface floaters.
About 10 to 13' long (3 to 4m).

Spanish Dancer
Hexabranchus sanquinensis
Bottom crawlers, but can also swim.
Grows to about 10" (25 cm).

Christmas Tree Worm
Spirobranchus giganteus
Coral reefs.
Grows to about 1 1/4" (3 cm).

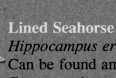

Lined Seahorse
Hippocampus erectus
Can be found among eelgrasses and other sea vegetation.
Grows to about 5" (13 cm).

Boxer Crab
Lybia tessellata
Coral reefs; carries two anemones it uses for "boxing"/protection.
Grows to about 3/4" (2 cm).

Fat Sleeper
Dormitator maculatus
Ponds, saltwater marshes, freshwater streams.
Grows to about 10" (25 cm).

Scrawled Cowfish
Lactophrys quadricornis
Usually in grassy areas in shallow water.
Grows to about 19" (48 cm).

Jellyfish
Cephea cephea
Freely floats offshore, open seas.
Grows to about 20" (50 cm).

Trumpetfish
Aulostomus maculatus
In shallow water around reefs.
Grows to about 30" (76 cm).

Clown Anemonefish
Amphiprion percula
Lives among anemones.
Grows to about 3 1/4" (8 cm).

Common Flyingfish
Exocoetus volitans
Surface dwellers.
Grows to about 12" (30 cm).

Sergeant Major
Abudefduf saxatilis
Reefs, grass beds.
Grows to about 7" (18 cm).

Naked Sole
Gymnachirus melas.
Sandy areas along the coast.
Grows to about 6 1/4" (16 cm).

Spiny Dogfish
Squalus acanthias
Off the coast over soft sea bottoms.
Grows to about 5' (1.5 m)

Longlure Frogfish
Antennarius multiocellatus
Reefs.
Adult size is from 4 1/2" to 6" (11-15 cm).

Footballer Trout
Plectropomus laevis
Reefs.
Grows to about 2.5" (6 cm).

Schoolmaster
Lutjanus apodus
Around rocks and coral reefs in coastal waters.
Grows to about 24" (61 cm).

Sailfish
Istiophorus platypterus
Open seas.
Grows to about 10' 9" (3.3 m).

References:

The Audubon Society Field Guide to North American Fishes, Whales & Dolphins, by:
    Atlantic and Gulf Coast Fishes: Herbert T. Boschung, Jr.
    Freshwater Fishes: James D. Williams
    Pacific Coast Fishes: Daniel W. Gotshall
    Whales and Dolphins: David K. Caldwell, and Melba C. Caldwell
    Visual Key: Carol Nehring and Jordan Verner

Peterson Field Guides, **Coral Reefs**, by Eugene H. Kaplan

Coral Reefs, Nature's Richest Realm, by Roger Steene

Coral Kingdoms, by Carl Roessler

Tropical Fishes of the Great Barrier Reef, by Tom C. Marshall

The Complete Divers' and Fishermen's Guide to FISHES OF THE GREAT BARRIER REEF AND CORAL SEA, by John E. Randall, Gerald R. Allen and Roger C. Steene

The Audubon Society Book of Marine Wildlife, by Les Line, with text by George Reiger

Pirate Perch
Aphredoderus sayanus
Streams, ponds, and swamps.
Grows to about 4 1/2" (11.5 cm).

SMS: For my parents, Richard and Iola Samson

TPN: For Aunt Hazel

Text Copyright © 1995 by Suzanne M. Samson
Illustrations Copyright © 1995 by Preston Neel
International Standard Book Number 1-57098-052-7
Library of Congress Catalog Card Number 95-69269

Published by
ROBERTS RINEHART PUBLISHERS
5455 Spine Road, Boulder, Colorado 80301

Published in the UK and Ireland by
ROBERTS RINEHART PUBLISHERS
Trinity House, Charleston Road
Dublin 6, Ireland

Distributed in the U.S. and Canada by Publishers Group West

Printed in Hong Kong